P9-CIU-785

California
Plants & Animals

Stephen Feinstein

Heinemann Library
Chicago, Illinois

a division of Reed Elsevier Inc.
Chicago, Illinois

Customer Service 888-454-2279

Visit our website at www.heinemannlibrary.com

Designed by Heinemann Library
Page layout by Depke Design
Printed and bound in the United States by
Lake Book Manufacturing, Inc.

07 06 05 04 03
10 9 8 7 6 5 4 3 2 1

Library of Congress Cataloging-in-Publication Data
Feinstein, Stephen.
California plants and animals / Stephen Feinstein.
p. cm. -- (State studies)
Includes bibliographical references (p.).
Contents: Wild California -- Native plants and animals -- Endangered species -- Extinct species -- Then and now. ISBN 1-40340-343-0 (hardcover) -- ISBN 1-40340-560-3 (pbk.)
1. Natural history--California--Juvenile literature. [1. Zoology California. 2. Botany--California. 3. Endangered species.]
I. Title. II. Series: State studies (Heinemann Library (Firm))
QH105.C2F45 2002
578'.09794--dc21

2002010885

Some words are shown in bold, **like this.** You can find out what they mean by looking in the glossary.

Acknowledgments
The author and publishers are grateful to the following for permission to reproduce copyright material:

Cover photographs by (top, L-R) Tom McHugh/Photo Researchers, Inc., Stephen Ingram/Animals Animals, Phil Degginger/Animals Animals, William Dow/Corbis, (main) Inga Spence/Visuals Unlimited

Title page (L-R) Adam Jones/Visuals Unlimited, Adam Jones/Photo Researchers, Inc., Joe McDonald/Visuals Unlimited; contents page (L-R) Tom McHugh/Photo Researchers, Inc., John Elk III/ELKJO/Bruce Coleman, Inc.; p. 4 Corbis; p. 5 Leonard L. T. Rhodes/Animals Animals; pp. 6, 8, 45 maps.com/Heinemann Library; p. 9T Dan Suzio/Photo Researchers, Inc.; p. 9B Robert Y. Eunice Pearcy/Animals Animals; p. 10 Harold and Judy Del Ponte; p. 11T James Blank/Bruce Coleman, Inc.; p. 11B Don Skillman/Animals Animals; p. 12T David Welling/Animals Animals; p. 12B Adam Jones/Photo Researchers, Inc.; pp. 13, 29, 40 Galen Rowell/Corbis; p. 14 S. Michael Bisceglie/Animals Animals; p. 15 John Elk III/ELKJO/Bruce Coleman, Inc.; pp. 16, 19B, 21B Thomas Hallstern/Outsight.com; pp. 17, 24 Randy Wells/Corbis; p. 18 Stephen Ingram/Animals Animals; p. 19T Ed Cooper; p. 20 Dennis Flaherty/Photo Researchers, Inc.; p. 21T Adam Jones/Visuals Unlimited; p. 22 Ralph A. Clevenger/Corbis; p. 23 Carl Roessler/Animals Animals; p. 25 Kevin Schafer/Corbis; p. 26 Gary Zahm/ZAHMG/Bruce Coleman, Inc.; p. 28 George D. Lepp/Corbis; p. 30 Robert C. Fields/Animals Animals; p. 31 Arthur Morris/Visuals Unlimited; p. 32 Tom McHugh/Photo Researchers, Inc.; p. 33 Ted Levin/Animals Animals; p. 34T Neil Rabinowitz/Corbis; p. 34B Bob Ecker/Heinemann Library; p. 35 Dr. David Schwimmer/Bruce Coleman, Inc.; p. 36 D. & R. Sullivan/Bruce Coleman, Inc.; p. 37 A. J. Copley/Visuals Unlimited; p. 38 Joe McDonald/Visuals Unlimited; p. 39 Historical Picture Archive/Corbis; p. 41 Steve Maslowski/Photo Researchers, Inc.; p. 42 Jim Balog/Photo Researchers, Inc.; p. 43 Mark E. Gibson/Visuals Unlimited; p. 44T John Bova/Photo Researchers, Inc.; p. 44B Steve Strickland/Visuals Unlimited

Photo research by Julie Laffin

Thanks to expert reader, John O. Sawyer, Professor Emeritus of Botany, Humboldt State University, for his help in the preparation of this book. Also, special thanks to Lucinda Surber for her curriculum guidance.

Every effort has been made to contact copyright holders of any material reproduced in this book. Any omissions will be rectified in subsequent printings if notice is given to the publisher.

Contents

Wild California

California has an amazing number of **species** of plants and animals. Scientists believe that the uniqueness of many California plants and animals is partly the result of California's geographical isolation from the rest of the country. The Sierra Nevada mountain range, with many peaks rising above 14,000 feet, along with other high mountain ranges to the north and east, form natural **barriers.** The vast deserts of southeastern California do also. The state's natural barriers have kept out many non-native species.

The Sierra Nevada mountain range contains over half the plant species, two-thirds of the birds and ***mammals,*** *and half the reptiles and amphibians found in California.*

Introduced Species

Not all of California's plant and animal species are native to the state. For example, in 1856, the eucalyptus tree was brought to California from Australia. The eucalyptus was planted along the California coast, in coastal mountains, and in the Central Valley, below 1,000 feet in elevation. It is used as a windbreak, to block noise or views, as an ornamental plant, or for firewood. Eucalyptus leaves have an unusual smell. The bark shreds in long strips each year, revealing smooth bluish-gray, tan, and whitish inner bark. Eucalyptus trees are now so common throughout California that it is easy to mistake them for native plants.

Among California's many native plants and animals are some that are unique. The coast redwood, California's official state tree, draws visitors from around the world. California's ancient **bristlecone pines** are the oldest living trees in the world. Many unusual plants, such as the Joshua tree, can be found in California's deserts. Other California plants, while not so unusual, stand out because they can be found in so many parts of the state. The oak tree and California poppy, for example, are standard features of the California landscape.

California Ecosystems

Not surprisingly, the great variety in California's geography has resulted in many different **ecosystems** throughout the state. An ecosystem includes all the

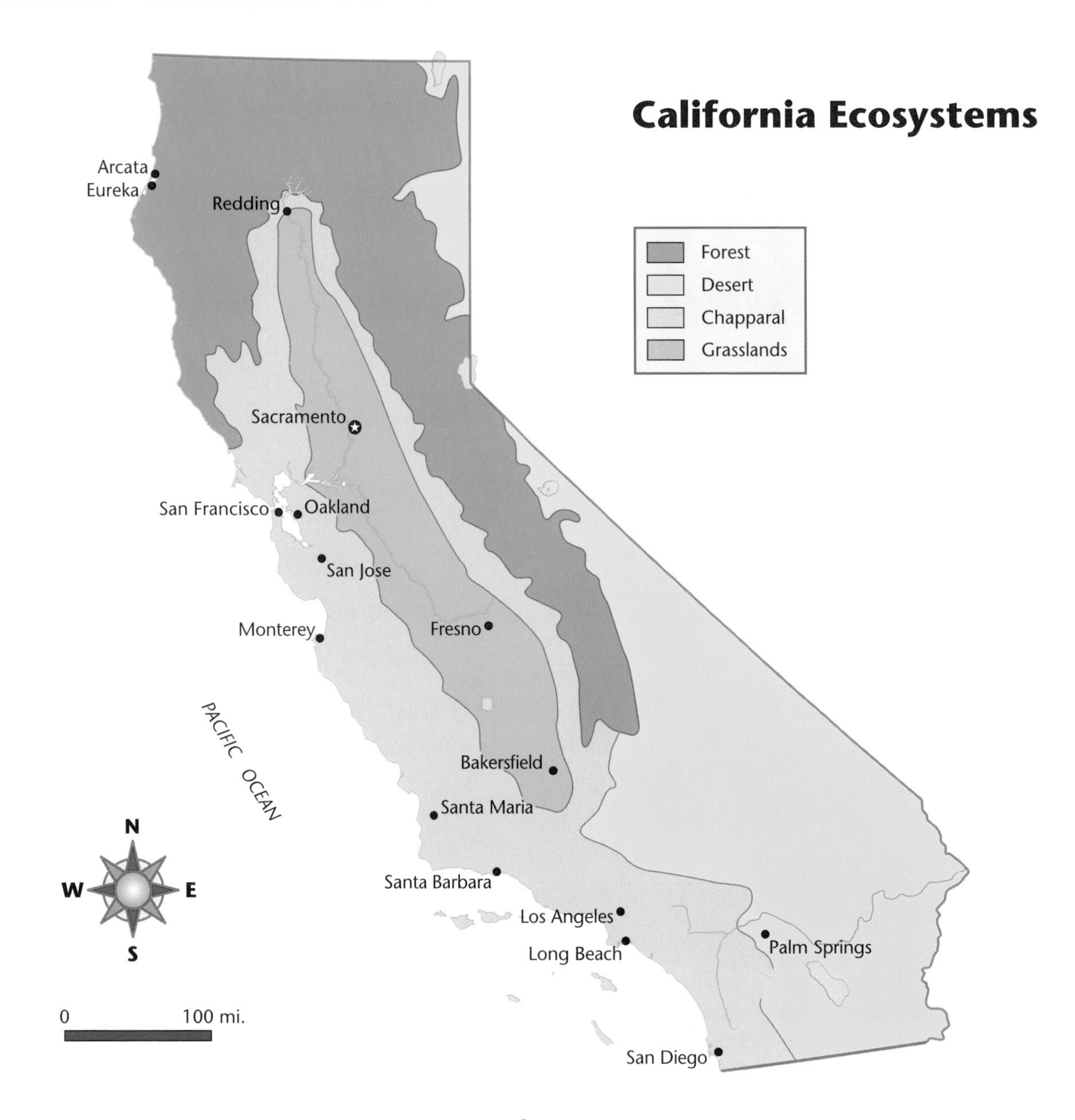

*California's different ecosystems support many types of plants and animals, many that are **endemic** to the state.*

communities of living things in an area, as well as nonliving factors such as the water, soil, and **climate.** All parts—the living **organisms** and the nonliving factors—exist in a delicate balance. The major kinds of **ecosystems** in California are forest, desert, **chaparral,** and grassland.

Forests

California has 58,281 square miles of forest. That's more than a third of the entire state. Nowhere in the United States is there a greater variety of trees and forests than in

The Food Cycle in an Ecosystem

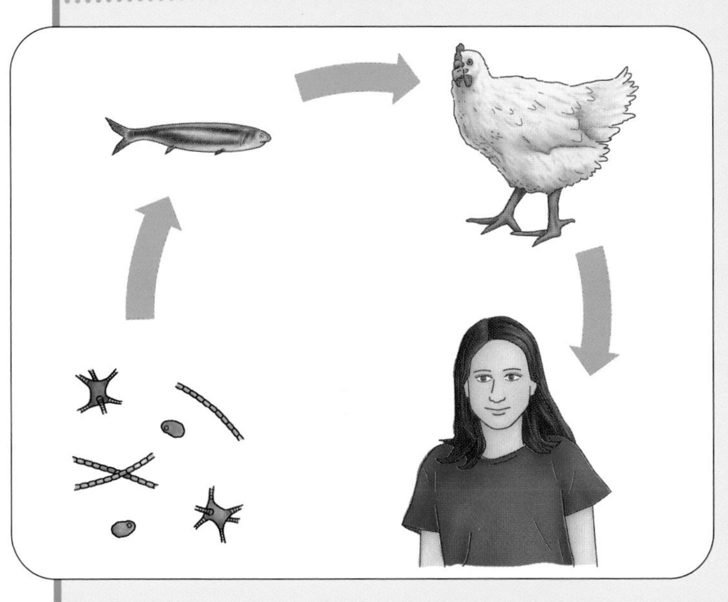

A balanced ecosystem supports and maintains the organisms in it through a process known as the food cycle. All organisms are **producers** or **consumers.** Green plants are producers. They make their own food through **photosynthesis.** Plant cells contain chloroplasts, which are tiny structures that convert sunlight, water, and carbon dioxide into sugars. The chloroplasts are green because they contain chlorophyll, which is a green pigment that absorbs the energy in sunlight.

Animals are consumers. They eat the food that the producers make. There is often more than one level of consumer in an ecosystem. For example, an anchovy eats plankton. The anchovy is then eaten by a chicken. The chicken is eaten by a human or an animal eating pet food. The anchovy is a first-level consumer, the chicken is a second-level consumer, and so on. Some consumers, such as bacteria and fungi, feed on dead organisms and are called **decomposers.** They break down dead matter, returning it to the soil as nutrients. Plants then take in the nutrients, and the food cycle goes on.

California. California forests are home to more than 4,000 **species** of plants and over 400 species of animals.

There are more than 50 species of trees in California. It is the only place in the United States to have coast redwoods and **giant sequoias.** On the north coast, the most

California wildlife is protected from hunters in many of the state's parklands.

common trees are Douglas fir and redwood. The central coast is home to redwoods. The northern interior and southern California forests are mostly a mix of fir and pine trees.

California forests experience great variations in temperature from season to season. This affects the behavior of animals living there. During the very hot summer days, **nocturnal** animals, such as raccoons, flying squirrels, bats, and opossums, sleep during the day and have an active life of hunting for food at night. Some animals, including songbirds and some butterflies, **migrate** south to warmer and sunnier places for the winter. Other forest animals survive the winter by hibernating, or sleeping.

Big Basin State Park in Boulder Creek is California's oldest state park, established in 1902. It is home to over 18,000 acres of both old-growth and new redwood trees.

Unusual California Trees

It may be hard to believe, but one tree in California is so big that a person can drive a car through the opening in its trunk. The tree is in northern California, about 200 miles north of San Francisco, on an old highway called Avenue of the Giants. The tree is a coast redwood **endemic** to the state. A large opening has been cut into the "Drive-Thru-Tree." Tourists line up and, one-by-one, slowly drive their cars through the tree. Amazingly, the tree is still alive today.

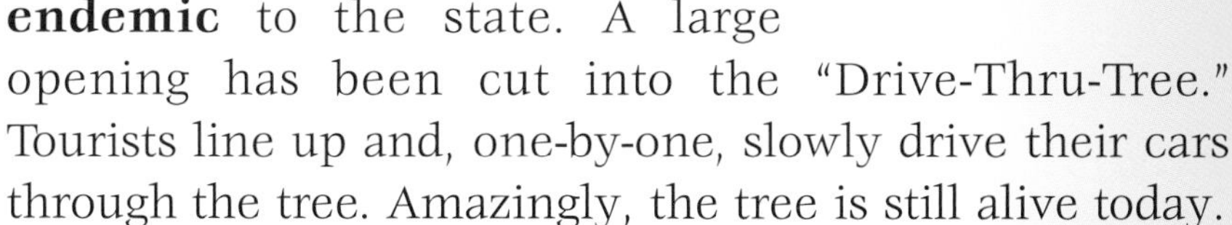

Many of California's plant ***species*** *only grow in California. For example, the Monterey Cypress is a tree that can only be found on the Monterey* ***peninsula*** *on the coast of central California.*

Putting a road through a tree could not happen as easily today, because laws protect the redwoods in many places in California.

Redwoods are **conifers.** This means they are trees with cones. True giants among trees, some coast redwoods are more than 360 feet tall. But the coast redwood's cousin in the Sierra Nevada, known as the **giant sequoia,** is even bigger. It is not quite as tall as the coast redwood, but it is much thicker. The General Sherman giant sequoia in Kings Canyon–Sequoia National Park is 275 feet tall, which is about as high as a 25-story building. It is 36.5 feet thick at its base, and 103 feet around.

Sixty million years ago, redwood trees grew throughout North America. They flourished for millions of years. Then, one million years ago, the thick **glaciers** of the **Ice Age** began creeping south across the continent. The redwoods of California were protected from the sheets of ice by the natural mountain **barriers** to the north and east. They were the only redwoods that survived. Many of today's redwoods are more than 1,250 years old, making them ancient compared to most other trees in California. That is because the bark of redwoods is highly resistant to the fires that occasionally sweep through California forests.

Visitors to a coast redwood forest often compare the experience to entering a large cathedral or other place of

The General Sherman tree is the largest living tree on Earth.

worship. It is cool, moist, and silent on the forest floor beneath the towering redwoods. The upper branches of the redwoods, often surrounded by fog, filter out much of the light. Water collects on the needle-like leaves and drips down the trunk into the ground, where it is absorbed by the shallow roots. Unfortunately, much of California's redwood forests have been cut down over the past 150 years by the lumber **industry.** Many of the remaining redwood groves are now protected from further logging.

Many other unusual trees can be found in California. The California bay tree, also known as the bay laurel, is the source of bay leaves. The shiny, leathery leaves give off a strong spicy scent when crushed. The Pacific

Earth's Oldest Living Trees

High in California's White Mountains, near the Nevada border, is the Ancient **Bristlecone Pine** Forest. Clinging to life in the cold, thin air, at elevations of 10,500 feet and above, are some of the oldest living things on Earth. Many of the knotted and twisted bristlecone pines are more than 4,000 years old! Amazingly, these trees, which were already ancient when the city of Rome was founded in Italy, are still alive! Over the millennia, the ancient trees have been blasted by bitter winds and twisted into odd shapes. No one knows how much longer they will live.

Native peoples used the bark of the Pacific madrone to make tea.

madrone has a smooth red-brown bark that peels away to show a yellowish layer beneath.

Deserts

California's 25,000 square miles of desert can be divided into two basic zones: the Mojave, or high desert, and the Colorado, or low desert. Both of these areas contain unique plants and animals, as well as natural **resources.** Most deserts are very hot and dry. In Death Valley, part of the Mojave Desert, daytime temperatures often reach over 100°F.

Although deserts are usually very hot places with little rain, desert plants have adapted to survive in the harsh growing conditions.

Many desert plants have thin, spiky leaves that prevent water loss. Others, like the Joshua tree, have needle-like leaves with a waxy coating that prevents

A common sight in the desert is the roadrunner. It races along the ground at up to fifteen miles per hour. Roadrunners are fast enough to catch rattlesnakes!

evaporation. Plants can also get water through deep roots. The mesquite tree has roots that can extend 100 feet into the ground, tapping water from underground supplies.

There is a variety of animals in the desert, including jack rabbits, bobcats, coyotes, desert tortoises, rattlesnakes, and tarantulas. They all have one thing in common: the need to conserve energy and water. Most desert animals are active during the morning, evening, and after dark, the coolest times of the day. Many desert animals are also light in color so they absorb less heat from the sun.

The Mojave Desert in southern California is home to the Joshua tree. This strange-looking plant can best be seen in Joshua Tree National Monument. The Joshua tree belongs to a family of plants known as lilies, and grows to a height of thirteen to sixteen feet. Joshua trees have arms that grow at odd angles and twist into almost humanlike gestures. The Joshua tree has a ring of spikes around the tip of each branch. Its dark green leaves are shaped like the point of a spear. They are stiff, leathery, and as sharp as a knife. The small size of

Joshua trees do not have growth rings like other trees, making it difficult to determine their age.

the leaves means the plant does not lose too much of the water stored in its roots. That is important in order to survive in the desert.

The Joshua tree is the center of a complex **habitat.** It is home to many different kinds of creatures. Living in and around a Joshua tree are woodpeckers and other birds, wood rats, termites, yucca moths, snakes, and lizards.

Growing near streams in canyons in the southern California deserts is the rare California washingtonia, also known as the California fan palm. This tall palm tree, **endemic** to California, has fan-shaped leaves that are three to six feet in diameter. The trunk may be covered with hanging dead leaves. Native Americans used to grind the seeds of the California washingtonia into flour.

Chaparral

Many different California plant communities are referred to as scrub, a type of vegetation dominated by shrubs. They consist of different groups of **species** according to the habitats. There are desert scrub, dune scrub, and coastal scrub communities. **Chaparral** is a common type of scrub community that grows at low elevations away from the coast.

Location plays an important role for the chaparral **ecosystem** in southern California. This ecosystem has a mild, rainy season during the winter. The hot, dry, summer season has average temperatures that range from 75° to 90°F. Hot, dry periods influence this ecosystem's plants, many of which are small shrubs and bushes. Most of the plants are less than ten feet high. In many regions, these shrubs are evergreen and typically have small, thick, waxy leaves designed to retain moisture. They are adapted to survive long periods of time with no rain. Herbs, such as sage, are also found here. Manzanita—which has smooth red bark similar to madrone—ceanothus, and chamise are typical chaparral plants.

You can see chaparral in the coastal mountains around San Diego, Los Angeles, Santa Barbara, San Luis Obispo, San Jose, and in the foothills of the Sierra Nevada and Klamath Mountains.

The chaparral also contains many different animals, including mountain lions, hawks, eagles,

The word chaparral *comes from the Spanish word* chapparo, *meaning "scrub oak."*

scrub jays, and roadrunners. Reptiles also live in the **chaparral.** Rattlesnakes are an important part of the chaparral because they keep rodent populations from getting out of control.

Grasslands

The Central Valley area of California is grassland, but this **ecosystem** is not as large as it once was. In the 1500s, Europeans arrived and brought with them seeds from their homeland, which had a similar **climate.** These new seeds soon swept across California, growing new plants that were disruptive to the native plants that were already living here. Since then, most native California grasslands have disappeared, due to farming and urban growth during the last 200 years.

Plants in temperate grasslands need to adapt to cold winters, hot summers, and drying winds. Since dryness is always a concern, plants have adapted ways to conserve water. One way they conserve water is by having thin, needle-like leaves that expose little of the plant to the sun. Grasses also have extensive roots that prevent grazing animals from pulling their roots out of the ground. Since grass grows upward from its base, it is less likely to be damaged by fires, animals, and humans than most other plants.

Most temperate grasslands are inhabited by rabbits, mice, and ground squirrels. These animals have sharp teeth that are easily able to gnaw through grass. Snakes, owls, foxes, deer, hawks, and coyotes can also be found here.

Variety of California's Wildlife

California contains the greatest variety of plant and animal life in the country. More than 5,000 types of plants, including trees, shrubs, flowers, grasses, ferns, and cacti, are native to California. About a third of California's plants cannot be found anywhere else in the world.

Efforts to preserve the **environment** have helped endangered animals made a dramatic comeback. These animals include elk, mountain lions, sea lions, white sharks, largemouth bass, pelicans, raccoons, cranes, sea otters, swans, wild pigs, and whales. Continued concern and support by California residents will ensure that more **species** are protected and saved in the future.

By law, mountain lions can be killed only when they pose a threat to people, pets, livestock or a person's property. They cannot be hunted for sport.

Native Plants and Animals

California's state flower is the California poppy, also known as the golden poppy. The poppy covers fields and rolling hills in the springtime with a golden-orange carpet. Like the oak tree, the poppy is so common in many parts of the state that it has become a regular feature of California landscapes.

Sailors called California the "land of fire" because of the many poppies growing on the foothills.

The California coast is home to many kinds of **succulents.** These plants store water inside enlarged cells. The plant's long roots reach water that has a high content of sea salts, located deep in the soil. By storing water in its tissues, the plant can lessen the concentration of salts. It can also survive long periods of no rain more easily than other types of plants. Many succulents are native California plants, such as owl's clover and Live-Forever. However, one of the most common succulents, the ice plant, originated in Africa. It has adapted so well to the California coastal **environment** that it has become a common sight up and down the coast.

The ice plant needs lots of sunshine to grow properly.

Growing in the **marshes** and **wetlands** of California's Central Valley is a plant known as **tule,** which resembles a cattail. California native peoples such as the Pomo and Yokut used tule as a building material. They also used long stalks of tule, bundled tightly together, as the frame for rafts and canoes.

*There are seventeen **species** of tule in California. One species—Common tule—is shown here.*

This blue oak can be found in the foothills of the Sierra Nevada mountains.

Oak Trees

California's oak trees can be found throughout the state, in many different plant communities. The acorn, produced in huge quantities by oak trees, was the most important food for California's Native Americans. There are sixteen **species** of oak in California. Eleven of them are **endemic** to the state. Some oaks are **deciduous,** meaning each year they lose all of their leaves at the same time. Oak trees can be found in the Coast Range, interior valleys, foothills of the Sierra Nevada, and on the islands off the coast of southern California. Oak trees can even be found in the desert. The most common types of oak trees include the valley oak, blue oak, coast live oak, interior live oak, and California black oak.

Notable California Animals

Mammals such as squirrels, rabbits, and raccoons, are very common throughout California. So are deer, whose population has grown to more than one million. Even black bears are fairly common in the mountains. Some California animals are especially rare, such as mountain lions. Mountain lions are also known as cougars, or pumas. In recent years, human homes in many places have expanded beyond the **suburbs** into areas that used to be wild. This has increased the chances of an encounter between a person and a mountain lion. On very rare occasions, such encounters end badly.

The howl of the coyote is known to some as the "song of the West."

Coyotes can often be seen slinking through the brush in many parts of California. The coyote, a member of the dog family, is well-adapted for survival in the mountains, desert, and even along the coast. The coyote is a scavenger—it eats whatever it can find. California's native peoples respect the coyote. According to a creation myth of the Yokut Indians, Coyote created human beings.

California Indians also admire the tule elk, referring to it as *wapiti*, meaning "white rump." Elk are the largest member of the deer family, and can weigh as much as

After elks drop their antlers in March, a new set begins growing immediately. Antlers can grow up to an inch a day!

1,200 pounds. That's as much as six to eight adult men would weigh! People searching for gold almost wiped out the tule elk in the 1840s. In 1885, there were only 28 left. This number has now increased to over 900, mainly because of three reserves in California's **chaparral** region. California elk are listed as endangered.

Marine Life

California's marine, or water, life has just as much variety as all other life-forms in the state. Along the coast, sea lions can be seen on the beaches and the piers in San Francisco and Monterey. At certain times of the year, **migrating** whales can be seen just offshore.

Great white sharks occasionally swim in the waters of the Gulf of the Farallones between San Francisco and the Farallon Islands, 25 miles offshore. During the fall, there are more great whites in this area than anywhere else in the world. The sharks are drawn here by the great numbers of sea lions, their favorite food. The great

*Great white sharks are a **protected species** along the coast of California.*

The Value of Anchovies

If you've ever eaten Caesar salad, you've probably eaten anchovy, the little six-inch fish that gives the salad its special flavor. Humans are not the only creatures that eat anchovies. The anchovy is a favorite food of many birds. And, the anchovy is the main source of food for many **marine mammals** and other fish. About 65 percent of the large anchovy population in the waters off the California coast become food for other **species** each year. The anchovy is also an extremely valuable **resource** for the fishing **industry.** Vast numbers are caught each year. Anchovies are packaged for human consumption or canned for pet food. Also, much of the catch is ground up into fish meal to feed **poultry** and livestock.

Although the anchovy is not currently considered an **endangered species,** the disappearance of the anchovy would put the entire Pacific marine food chain at risk. So, the government has limited the number of anchovies that can be caught each year. Also, the government has established a reserve of this fish to make sure that it continues to exist.

whites can grow up to 15 feet in length and weigh as much as 3,000 pounds. On rare occasions, they attack humans, possibly mistaking them for sea lions. Surfers unlucky enough to be in the wrong place at the wrong time have been seriously injured or even killed by great white sharks.

The Gulf of the Farallones is also the place to find salmon, which feed on the abundant supply of anchovies, herring, sardines, squid, and shrimp. Salmon are born in California

Making the trip home to spawn is a long, difficult journey for salmon.

rivers, but then spend most of their life in salt water. They often roam as far as 1,800 miles out to sea before returning to their place of birth to **spawn.** On their return, the salmon have to swim upstream, battling the river currents. For thousands of years, salmon was an important part of the California native peoples' diet.

Birds of California

California is home to over 600 **species** of birds. This is nearly two-thirds of all birds found in the U.S. Some of the birds in California are quite common, such as the seagull and the band-tailed pigeon. Others are quite rare, including the California condor, the snowy owl, and the bald eagle.

The Western gull is the most common of all the sea birds in California. In the summer, visitors taking a boat from San Francisco to the Farallon Islands can see thousands of gulls, murres, comorants, and puffins. This is the largest colony of nesting seabirds in the continental United States. The birds flock to these small rocky islands to breed. By fall, they have all flown south for the winter.

Scientists estimate there are 400,000 birds on the Farallon Islands.

The California valley quail is a common sight in California, as well as the state bird. Also common throughout the state's oak woodlands is the acorn woodpecker. It is easily spotted by its bright red head and the distinctive sound of its pecking at dead trees, creating spaces to store the acorns and insects it has gathered for food.

Crows are very common in California. Less common is the raven, a larger version of the crow. Glossy black ravens can be seen in the mountains, gliding silently upward on rising **thermals,** which are heated columns of air. Various groups of California's native peoples regarded the raven as the creator of the world.

Endangered Species

A huge increase in California's human population began in 1849, with the gold rush. This resulted in a huge decrease in the populations of many of California's animal **species.** By 1900, many species had practically disappeared, due to a combination of hunting, trapping, poisoning, and loss of **habitat.** As a result, some animals were designated by the government as **endangered, threatened,** or **protected species.** Among the most seriously endangered species were the California condor, bald eagle, golden eagle, peregrine falcon, hawk, heron, brown pelican, sandhill crane, swan, beaver, fox, mink, elk, sea otter, sea lion, and whale.

Sea otters sleep and eat while on their backs, and usually swim on their backs, too. California sea otters spend almost all of their time in the water.

Endangered and Threatened Species

The California Endangered Species Act allows individuals, groups, and the Department of Fish and Game to recommend animals and plants to be classified as threatened or endangered. These recommendations come when the **environment** of the plant or animal has been altered in some way, such as by competition, disease, natural occurrences, or overcollection by humans.

Threatened in California

Alameda whipsnake
Bank swallow
Barefoot banded gecko
California black rail
Coachella Valley fringe-toed lizard
Coastal California gnatcatcher
Desert tortoise
Flat-tailed horned lizard
Giant garter snake
Greater sandhill crane
Green sea turtle
Guadalupe fur seal
Island fox
Island night lizard
Loggerhead sea turtle
Mohave ground squirrel
Mountain plover
Northern spotted owl
Olive (Pacific) Ridley sea turtle
Peninsular bighorn sheep
San Clemente sage sparrow
San Joaquin antelope squirrel
San Joaquin kit fox
Sierra Nevada red fox
Southern rubber boa
Southern sea otter
Stellar (northern) sea lion
Stephens' kangaroo rat
Swainson's hawk
Western snowy plover
Wolverine
Yuma clapper rail

Endangered in California

Amargosa vole
Arizona Bell's vireo
Bald eagle
Belding's savannah sparrow
Blue whale
Blunt-nosed leopard lizard
Brown pelican
Buena Vista Lake shrew
California (Sierra Nevada) bighorn sheep
California clapper rail
California condor
California least tern
Elf owl
Fin whale
Fresno kangaroo rat
Giant kangaroo rat
Gila woodpecker
Gilded northern flicker
Great gray owl
Humpback whale
Inyo California towhee
Least Bell's vireo
Leatherback sea turtle
Light-footed clapper rail
Marbled murrelet
Morrow Bay kangaroo rat
Pacific pocket mouse
Point Arena mountain beaver
Right whale
Riparian brush rabbit
Riparian woodrat
Salt-marsh harvest mouse
San Bernardino kangaroo rat
San Clemente loggerhead snake
San Francisco garter snake
San Miguel Island Fox
Santa Catalina Island Fox
Santa Cruz Island Fox
Santa Rosa Island Fox
Sei whale
Short-tailed albatross
Southwestern willow flycatcher
Sperm whale
Tipton kangaroo rat
Western yellow-billed cuckoo
Willow flycatcher

Early Government Actions to Protect Some Species

During the 1900s, a growing number of scientists became alarmed at the shrinking populations of animal **species.** The sea otter, for example, had been hunted almost to **extinction** during the 1800s because of its valuable fur. As the general public became aware of the problem, there was growing support for government action. In 1911, the Fur Seal Treaty, signed by the United States, Russia, Japan, and Great Britain, gave some protection to the sea otter in the Pacific Ocean. In 1913, a California state law gave additional protection to the sea otter. But it seemed that there were hardly any otters left in California. Then, in 1938, a group of 100 otters was discovered along the California coast.

Golden Trout

The golden trout is **endemic** to California. It lives in the icy waters of mountain streams in the high Sierra Nevada mountains. The California State **Legislature** named the golden trout as the official state fish in 1947. Unfortunately, the numbers of golden trout grew smaller and smaller with each passing year. So in 1977, the U.S. Congress established the Golden Trout Wilderness Area to protect the native **habitat** of the golden trout. In 1978, the federal government added the golden trout to the list of **threatened species.** Thankfully, the golden trout are now growing in number in their mountain habitat.

In 1941, the California Sea Otter Game Refuge was established. By 1976, there were 1,700 otters. The federal government listed the sea otter as a threatened species in 1977. Now there are more than 2,000 sea otters in California. They mostly live off the coast of Monterey and Big Sur. The California sea otter is still protected as a threatened species. Other **marine mammals** have also benefited from protective laws, including sea lions, seals, and whales.

The Banning of DDT

During the 1960s, laws were passed banning the use of **DDT,** a **pesticide** that was widely used in the 1940s and 1950s in agriculture. During the 1950s, scientists had noticed that the bald eagle population was dropping quickly. They studied the problem and learned that DDT was the cause.

A normal peregrine falcon egg can be compared to one thinned by DDT's effects.

After DDT was sprayed on land and crops, rain washed it into streams. Fish in the streams were not killed by DDT, but they stored the pesticide in their bodies. When bald eagles ate these fish, they took in the DDT. The eagles did not die from the DDT, but they stored it in their bodies. Then, the eggs of eagles that had eaten fish containing DDT cracked during **incubation.** This prevented the development of baby eagles, so the population of bald eagles dropped.

The Endangered Species Act

In 1973, the federal government passed the **Endangered Species** Act and began making an official list of threatened and endangered species. An endangered species is one that is likely to completely die out soon if its situation is not improved. A threatened species is one that could become endangered in the near future. By 1985, the list of endangered species in the United States had grown to 350 **species.** Another 4,000 species were thought to be in trouble. Once a species was included on the list, it became a **protected species.** Thanks to public concern and government action, many of California's **endangered species** have made a comeback.

One of the species on the government list was the bald eagle, which is the national bird of the United States. The banning of **DDT** was one step in the right direction to help save the bald eagle. But there were other problems, such as the removal of nest trees and the shooting of eagles. By 1974, only 21 pairs of bald eagles were known to nest in California. Because of the protection of the government law, there are now more than 100 nesting pairs of bald eagles. In 1994, the bald eagle's status was changed from endangered to

In California, bald eagles are permanent residents in the north, and usually stay through the winter in the south.

threatened. There is now a proposal to remove the bald eagle from the list of **threatened species.**

A pelican's pouch underneath its bill can hold three times as much as its stomach.

Other birds harmed by the use of DDT were the brown pelican and peregrine falcon. By the 1970s, the population of brown pelicans in California was falling quickly. In 1970, on Anacapa Island off the coast of California, only one baby pelican was successfully hatched in more than 500 nesting attempts. Since then, however, the pelicans are back in large numbers. They can often be seen fishing for their dinner in waters up and down the entire California coast. Meanwhile, the peregrine falcon population increased from only 39 breeding pairs in all of North America in 1970, to 1,650 pairs in the year 2000, including 167 pairs in California.

California Condor

The California condor is another bird that was in danger of becoming **extinct.** At one time, there were thousands of condors in California. Today, the condor is the rarest bird in all of North America. Most of the condors were killed by settlers in the 1800s. Throughout the 1900s, even before condors were listed among the endangered species, there were laws in California against killing them. In 1947, the Sespe Condor Sanctuary was created in the mountains of southern California as a home for the birds. Even so, the population of condors fell until there were only a few birds left. The condors could not survive poisons in the **environment** and accidents, such as flying into power lines.

Condors can be seen at Condor Ridge, a large exhibit at the San Diego Zoo.

By 1979, there were fewer than 35 condors in the wild and only one in **captivity.** The following year, scientists began a major program to try to save the species. During the 1980s, each of the remaining 27 wild condors were caught and taken to the San Diego Wild Animal Park and Los Angeles Zoo. In 1988, the first condor chick was hatched in the San Diego Wild Animal Park. Others followed. In 1992, scientists released two of the condors that had been raised in captivity. They were taken to the mountains in southern California, which had been the **habitat** of their parents. By the year 2000, more than 50 condors had been released into the wild. Another 100 condors were still in zoos, but more would be released. There is reason to hope that the California condor has been saved from **extinction.**

Desert Tortoise

Federal laws also offer protection to endangered reptiles. The California desert tortoise is the only tortoise native to California, and it is the state reptile. It lives in the deserts of southern California. This gentle creature usually lives between 50 and 100 years. One tortoise in captivity lived to be 152 years old! The tortoise is amazingly well-adapted to its desert environment. But, once a tortoise is removed from its native habitat, it cannot survive for long without proper care. In the past, when people came across a tortoise in the desert, they would often take it home as a pet. Some

Adult desert tortoises are able to live in the desert because they can survive for over a year without water.

would sell the tortoise to a pet shop. Unfortunately, the tortoise would often die from lack of care. Some people were also hunting and killing tortoises in the desert.

The population of the desert tortoise has dropped about 90 percent in the last 50 years. The tortoise was listed as a **threatened species** under federal and state law. It is now illegal to remove a tortoise from its habitat, to harm it in any way, or to do damage to the habitat.

Marine Mammals Coming Back

The California gray whale, the official state **marine mammal,** can be seen all along the California coast from November to January, and again from March to May. The whales travel close to shore, usually in groups, or pods, of two to five. The gray whale can be 35–50 feet in length, weighing from 25–35 tons—about as big as a school bus.

Although the California gray whale was once hunted almost to extinction, it was given complete protection in 1938 by an international treaty. In 1994, the gray whale was considered a recovered **species.** Today, there are about 24,000 California gray whales. Other whales, such

Gray whales make an annual roundtrip of 13,000 miles from the cold Arctic waters along the Alaskan coast to the warm waters of Baja California in Mexico. This is the farthest ***migration*** *of any mammal.*

as the blue whale and the humpback whale, are still on the **endangered species** list.

Also back from near **extinction** are the California sea lion, Steller's sea lion, and the elephant seal. California and Steller's sea lions can be seen on Seal Rock in the ocean near San Francisco's Cliff House. Large groups of California sea lions can be seen at Pier 39 in San Francisco and Fisherman's Wharf in Monterey. Crowds of tourists are drawn to the big, sleek **marine mammals,** and can't resist feeding them. California sea lions can eat up to 15 to 20 pounds of fish a day, and a male can weigh as much as 900 pounds. Elephant seals can best be seen at Año Nuevo State Reserve, between Santa Cruz and San Francisco.

The large number of sea lions at Pier 39's K dock in San Francisco is a good example of humans and wildlife living together.

Extinct Species

The biggest threat to California wildlife today is the destruction of **habitat.** When an animal's habitat is destroyed, the animal **species** can die out. **Extinct species** are living things that existed in the past but no longer live. A species can be completely extinct, or extinct just in a certain area. For example, wolves no longer exist in California because they were hunted to extinction here.

Like many other parts of the world, California contains evidence of the creatures that lived there many thousands, even millions, of years ago. Scientists have discovered **fossils** of the trilobite in California. The trilobite was a creeping, crablike creature that lived in the sea. The oldest trilobite fossils are 600 million years old. The trilobite has been extinct for at least 300 million years!

Trilobite *means "three lobes." This name refers to the three sections of a trilobite's body.*

The La Brea Tar Pits

Not far from downtown Los Angeles is an unusual place known as the La Brea Tar Pits. At this spot about 40,000 years ago, during the last **Ice Age,** oil began oozing up from deep underground. When oil rose to the surface and mixed with oxygen, it thickened and formed a sticky black tar. When it rained, the water collected on top of the tar. A series of ponds or water holes were formed. Wild animals in the area did not realize that each water hole was really a trap—a tar pit. When the

Visiting the La Brea Tar Pits

In 1802, Jose Longinas Martinez, a Spanish traveler in southern California, wrote this description of the La Brea Tar Pits in his journal: ". . . there is a great lake of pitch, with many pools in which bubbles or blisters are continually forming and exploding. . . . In hot weather, animals have been seen to sink in it and when they tried to escape they were unable to do so, because their feet were stuck, and the lake swallowed them. After many years their bones have come up through the holes, as if petrified. I have brought away several specimens."

Scientists began digging and uncovering the La Brea fossils in 1906. Today, they are still finding, cleaning, identifying, and classifying the fossils from the tar pits. Visitors to the La Brea Tar Pits in Hancock Park, Los Angeles, can view life-size copies of the animals trapped in the tar. Reconstructed skeletons are also on display at the George C. Page Museum of La Brea Discoveries.

animals went into the water to bathe and drink, their feet got stuck in the tar at the bottom. The animals were trapped and remained there until they died.

Eventually the animal bones became **fossilized** in the tar pits. For about 20,000 years, animals continued to gather at the tar pits, becoming trapped. The resulting collection of fossils is the richest deposit of Ice Age fossils in the world.

Many of the animals that roamed California during the Ice Age became **extinct** after the Ice Age ended. Many of these **prehistoric mammals** were quite large. Among them were the **mammoth** and **mastodon,** relatives of today's elephant. There was also the saber-toothed cat. This huge cat, also known as the saber-toothed tiger, has become California's official state fossil. Other animals of that time included a giant wolf known as the dire wolf, a giant ground sloth, a bear that was much larger than any of today's bears, an early type of horse, an ancient North American camel, and a giant vulture called teratornis. All were trapped in the tar, and all became extinct. Also found in the tar

There is no animal alive today that is similar to the extinct two-hoofed mammal, the sloth.

were the fossils of many animals that still exist, such as rabbits and deer.

Why did so many animals become **extinct** after the end of the last **Ice Age?** At about that time, 8000 to 10,000 B.C.E., the first groups of California native peoples arrived in the area. They were **nomadic** hunters, and probably killed huge numbers of the animals. Also, the **climate** was changing and many animals could not adapt. The change in climate resulted in changes in the vegetation. When the plants that were a particular animal's main source of food disappeared, the animal died out.

Grizzly bears wait by streams for fish to swim by, then catch their meal.

The Grizzly Bear

In 1953, the **legislature** of California named the grizzly bear as the official state animal. This might seem like an unusual decision, because at the time there were no longer any grizzly bears in California. In fact, the last California grizzly died in 1922, when it was shot in Fresno County. But the grizzly is regarded as an important part of the state's **heritage.** At one time, there had been large numbers of grizzlies. They mainly lived in the San Francisco Bay Area, the Los Angeles foothills, the Central Valley, and the Sierra foothills. Grizzlies were the largest animals in California. A male grizzly can weigh up to 1,200 pounds—almost as

much as a small car. They were not threatened by any other animal except humans. For thousands of years, there were not enough humans around to threaten their survival. But then came the gold rush of the late 1840s, and the arrival of huge numbers of settlers, miners, and ranchers. This put the grizzly in danger.

Because the grizzly was not afraid of other animals, it was not likely to run from anyone. When meeting up with a hunter, it would stand its ground in challenge.

The grizzly bear's **habitat** was destroyed as human settlement expanded. Sheep ranchers shot grizzly bears because the grizzlies saw the sheep as a source of food. Sometimes ranchers poisoned the bears. Hunters shot grizzlies for sport as well as to sell the meat. The grizzly bear's behavior made the hunters' job easy, and the grizzly population in California fell quickly. Soon there were none. So, although grizzly bears still live in the mountains of Montana and some other wilderness areas, the grizzly became extinct in California.

Then and Now

A long list of animals were in danger of disappearing in California about 100 years ago. Thanks to efforts for the protection and preservation of **species,** many animals are back, sometimes in even greater numbers than before. Pelicans and wild sea lions can be seen in San Francisco. Whales, elk, coyote, and deer are just a few of the species that can be seen within 30 miles of the city.

Moving to the City

As humans build homes across more of the California countryside, less food is available here for the wild animals. All sorts of animals have left their usual **habitats** and **migrated** to the cities. Typical city birds, such as pigeons, have been joined by peregrine falcons and barn owls.

The peregrine falcon, an **endangered species,** has made a remarkable comeback in California. The fastest creature alive, the peregrine falcon can fly up to 200 miles per hour when hunting prey. In recent years, peregrine falcons have been building nests on ledges on the upper floors

Peregrine falcons have had to adjust to living in busy cities like Los Angeles.

of high-rise office buildings. Falcons swoop through the spaces between tall buildings to snatch up pigeons, as if they were flying through a canyon. Some falcons have even set up homes on the tops of bridge towers. The falcons sit there and watch for prey to pass by down below.

In many cities, raccoons can be found knocking over garbage cans at night. The fearless, **nocturnal** creatures are searching for food. Because they can eat almost anything, raccoons are well suited for city living.

City garbage cans are tempting for families of raccoons looking for food.

Black Bears and People

There are about 20,000 to 24,000 black bears in California. They are not a **threatened** or endangered species. Their habitat extends throughout all the mountain areas of California. The biggest black bear on record weighed 690 pounds, compared to a typical 1,200-pound grizzly. Also, unlike the grizzly, black bears have learned how to live together with humans. This is good for the black bear, because nearly 12 million hikers, campers, and fishermen spend time each year in the bear's **habitat.**

Unfortunately, many bears have developed a taste for human food. Once this happens, the bear will stop at nothing to reach the food. This includes ripping up a tent, smashing the windows of a car, or even entering a house. Campers should know that they must never, ever, feed a bear. They should also never keep food in or near their tent.

Black bears are not shy. This bear is looking for food that the campers may have in their tent.

How to Make a Bear-Proof Food Hang

The best way to keep food away from black bears is to hang the food in double-wrapped, heavy-duty plastic garbage bags. This can be done in two ways:

1) Tie a rock to a rope.
Throw the rock over a high, sturdy tree limb.
Tie a food bag to the rope and hoist it in the air.
Tie the end of the rope to another tree.

OR

2) Divide the food into two separate bags of equal weight.
Tie a rock to a rope.
Throw the rock over a high, sturdy tree limb.
Tie the food bag to the rope and hoist it in the air.
With the other loose end of the rope, tie off the other food bag while leaving a large fixed loop at the knot.
Using a long stick, lift the bag with the fixed loop.
As you raise the bag, the other bag will go lower.
Raise the bag so that both are high in the air, balancing one another, and well out of reach of bears.

Facing an Uncertain Future

The various attempts to protect California's animals have been very successful in many cases. But the future of some of the **threatened** or **endangered species** is still in doubt. **Environmental** damage or **habitat** destruction has a harmful effect. Offshore drilling and tanker transport of oil along the California coast sometimes results in oil spills in coastal waters. This often results in death or injury to countless sea birds, and **marine mammals** such as the sea otter. The oil ruins the otter's fur, leaving the animal with no protection against the cold.

*Oil spills can be devastating to marine and **mammal** life.*

There are other problems, too. For example, the population of the California mountain bighorn sheep is dropping quickly. There are very few of these sheep left in their high Sierra **habitat.** Although the federal government protects them as an **endangered species,** the government also allows **domestic** sheep to graze in the wilderness. Many bighorn sheep were wiped out by a disease introduced by the domestic sheep. Mountain bighorn sheep are often killed by mountain lions, which are also protected by the government. Since the mountain lion is a **protected species,** hunting them to save the sheep cannot happen either. Fish populations are threatened by huge water projects, such as the building of dams, and too much fishing.

Still, the attempts to preserve California's wildlife have been a success. Some **threatened** or endangered species have even been removed from the list of endangered species. If current efforts to protect California's endangered species and preserve the **environment** continue, California's **diverse** plants and animals will be around for a long time to come.

Desert bighorn sheep do not require a source for drinking water when there are green plants available.

Map of California

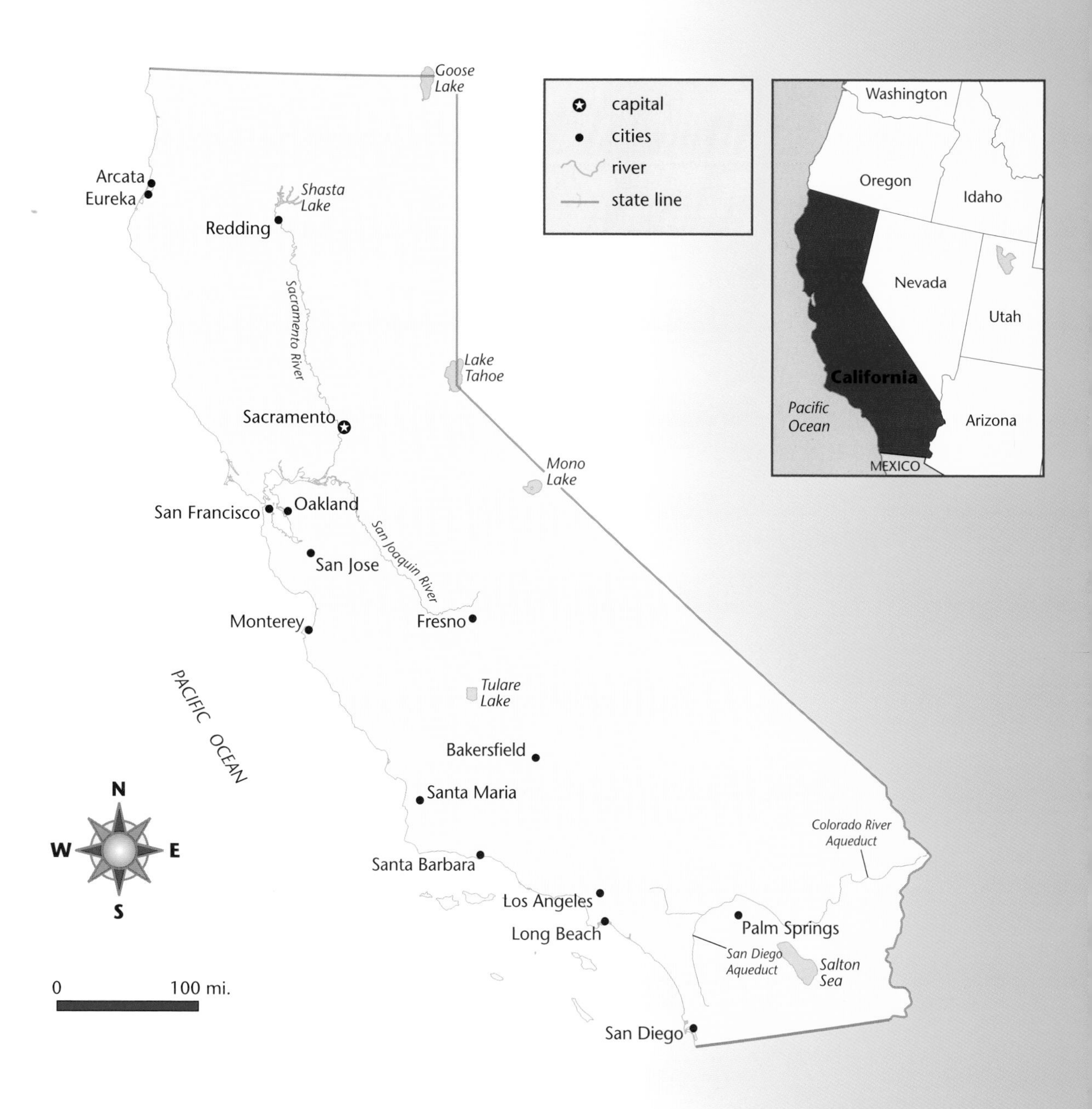

Glossary

barrier something that blocks the way

bristlecone pine oldest living tree on Earth

captivity under control of humans, often in a zoo

chaparral common type of scrub community that grows at low elevations away from the immediate coast in California

climate weather conditions that are usual for a place

conifer tree with cones

consumer organism in an ecosystem that eats the food the producers make. Some consumers eat other consumers.

DDT pesticide widely used in the 1940s and 1950s in agriculture

deciduous losing leaves each year

decomposer consumer in an ecosystem such as bacteria or fungi that feeds on dead organisms. Decomposers break down dead matter and return it to the soil as nutrients.

diverse having variety

domestic under the care of human beings

ecosystem community of living things, together with the environment in which they live

endangered species group of animals threatened with extinction

endemic growing or living only in a particular area

environment surrounding conditions that influence the life of plants and animals

extinct/extinct species living things that existed in the past, but are no longer alive

fossils/fossilized remains of prehistoric plants or animals that have turned to stone

giant sequoia largest living tree on Earth

glacier large sheet of ice that spreads or retreats very slowly over land

habitat conditions where an animal or plant lives and grows

heritage something that comes from one's ancestors

Ice Age period of time when a large part of the earth was covered with huge sheets of ice (glaciers) and the temperatures were cooler

incubation keep under conditions right for hatching or development

industry kind of business

legislature governmental body that makes and changes laws

mammal class of animals covered with hair (or fur or wool) that feed milk to their young

mammoth prehistoric relative of today's elephant

marine mammal aquatic mammal that lives in salt water, such as a sea lion, sea otter, or whale

marsh wet, low-lying area, often thick with grasses

mastodon prehistoric relative of today's elephant

migrate to move from one place to another for food or to breed

nocturnal active at night

nomadic not living in one location, but moving around to be close to a supply of food

organism living person, animal, or plant

peninsula piece of land extending over a body of water

pesticide substance used to destroy pests, such as DDT

photosynthesis process by which plants make food, using sunlight

poultry birds such as chickens, ducks, turkeys, and geese, grown to provide eggs or meat

prehistoric from the time before history was written

producers organisms in an ecosystem that make their own food from sunlight, water, and carbon dioxide, such as green plants

protected species group of animals protected by state or federal law because they are either endangered or threatened

resource valuable thing that can be made useful; there are natural and man-made resources

spawn produce or deposit eggs

species group of organisms that share the same physical characteristics

suburb city or town just outside a larger city; suburban means having to do with a suburb

succulent cactus or other plant that stores water in enlarged cells

thermal rising column of heated air

threatened species group of animals whose numbers are decreasing, bringing the group close to endangerment

tule rush with round stems; grows in wetlands and marshes of California's Central Valley

wetland very wet, low-lying area

More Books to Read

FitzGerald, Dawn. *Butterfly Hill: Saving the Redwoods*. Brookfield, Conn.: Millbrook Press, 2002.

Harder, Dan. *A Child's California*. Portland: WestWinds Press, 2000.

Kennedy, Teresa. *California*. Danbury, Conn.: Children's Press, 2001.

Pelta, Kathy. *California*. Minneapolis: Lerner Publications Company, 2001.

Povey, Karen D. *The Condor*. San Diego: Lucent Books, 2001.

Index

About the Author

Stephen Feinstein is a writer of educational materials, specializing in social studies and language arts. His extensive experience in observing and writing about historical, geographical, political, and cultural trends has given him special insights into creating concise yet informative presentations of complex subjects. He lives in Marin County, north of San Francisco, California.